OLDER DOG? NO WORRIES!

Maintaining physical, mental and emotional wellb... ...ldie

Sian Ryan

Hubble & Hattie

The Hubble & Hattie imprint was launched in 2009, and is named in memory of two very special Westie sisters owned by Veloce's proprietors.

Since the first book, many more have been added to the list, all with the same underlying objective: to be of real benefit to the species they cover, at the same time promoting compassion, understanding and respect between all animals (including human ones!)

Hubble & Hattie is the home of a range of books that cover all-things animal, produced to the same high quality of content and presentation as our motoring books, and offering the same great value for money.

www.hubbleandhattie.com

Disclaimer

Please note that no dog was deliberately frightened during photographic sessions. Any images used to depict dogs feeling worried about specific situations were taken whilst the animals were exploring the novel environment of the studio setup. These images were later modified to enable us to illustrate the points we wished to make. Please also note that all body language is context-specific, and individual signals can mean different things in different contexts.

Published in January 2020 by Veloce Publishing Limited, Veloce House, Parkway Farm Business Park, Middle Farm Way, Poundbury, Dorchester, Dorset, DT1 3AR, England. Fax 01305 250479/e-mail info@hubbleandhattie.com/web www.hubbleandhattie.com. ISBN: 978-1-787113-66-4 UPC: 6-36847-01366-0.
© Sian Ryan and Veloce Publishing Ltd 2020

Contents

Dedication & Acknowledgements

FOR COOPER AND RILEY

We shared so many adventures

Thank you

Thanks also to Calum Mcconnell and Graham Fisher for their expert photography

I take responsibility for the less professional shots

About the author

Sian Ryan gained her MSc in Clinical Animal Behaviour from the University of Lincoln with distinction in 2011. Whilst finishing her dissertation on Self Control in Pet Dogs she worked as a behaviour counsellor and trainer in the Lincoln Animal Behaviour Clinic, and went on to work as a researcher looking at novel ways of measuring emotions in dogs in 2012. With several years of dog training experience, Sian was the first course tutor for Life Skills for Puppies training classes and helped create and develop the course, as well as tutoring on the Life Skills for Puppies Trainers Courses offered by the University of Lincoln. Sian has been featured in all the daily national

newspapers and on local and national radio as a dog training expert, and her book *No Walks? No Worries!* was published in October 2014. Sian has also appeared as an animal training and behaviour expert for various BBC programmes, including Head Trainer for BBC2's acclaimed *Me and My Dog* (2017) and *Bang Goes the Theory*, and was consulted during the making of *Inside the Animal Mind* with Chris Packham.

Sian owns and runs the Developing Dogs Training Centre and Holiday Cottages in Cambridgeshire, and gives seminars and workshops nationally and internationally. Her speciality is pet dogs: developing interactions that enable dogs and owners to make the most of their unique relationship.

4

Introduction

I want my dogs to live forever, and want to enjoy their different life stages as much as possible, but I am well aware that living with an oldie can be upsetting, as the cost, time required, and emotional stress take their toll. It's true that older dogs can have more complex needs, require more vet trips and/or medication, and can no longer participate in the long walks of their youth, but sharing your home and heart with an older dog brings a whole new element to being a dog owner. It can be immensely rewarding, and also deeply upsetting. Ultimately, there is only going to be one outcome, but with preparation and flexibility we can ensure that those last days, weeks, months, and years are the best they can possibly be for everyone.

This book focuses on maintaining or even increasing quality of life for your older dog, and is written from the viewpoint that quality is more important than length of life. The aim is to exceed our dog's needs while ensuring we are also able to be objective over the impact on our lives, and realistic about what we can do to help. I am a firm believer that whenever it's safe or possible, the usual rules don't apply to older dogs. As a result my old lady is allowed, and positively encouraged, to be the pre-wash cycle on the dishwasher. If she nudges my knee while I'm eating my dinner, something will invariably fall from my plate into her mouth. If she wants some attention while I'm working, I'll do my best to give it to her. These moments of her personality showing through the old age are precious, and should be encouraged. She gets what she wants; the other dogs will get their turn in time.

Is my dog an older dog?

It seems like only yesterday that you took home a wriggly puppy or an active adolescent, but all too quickly our dogs age before our eyes. Sometimes this is a gradual decline that we barely notice, until one day a photo reminds us of how they were in

Bacon and Jo sharing a moment of mutual affection.

OLDER DOG? NO WORRIES!

Rules don't apply when you're 13 years old: Riley is allowed to pre-wash the dishes.

Plenty of life in the old dog yet. Freddie enjoying his trip to the beach.

their youth; at other times it can be a specific illness or injury that ages them apparently overnight.

Officially, the UK Kennel Club classes a dog as a veteran at 7 years of age; however, breed and size have an impact on whether a dog is truly a senior at what, for many dogs, is a comparatively young age. In general, the smaller the dog the longer they tend to live. Giant breeds, such as Great Dane or St Bernard, tend to have the shortest lifespan of between 7 and 10 years, whereas smaller dogs, like Jack Russell Terriers or Miniature Poodles, can live to 18 years or more. A reasonable measure is to regard your dog as an older dog when he reaches 75% of average lifespan of his breed: eg the average lifespan of a Staffordshire Bull Terrier is 12 to 14 years, so would be considered an older dog when he reaches 8 to10 years of age.

If your dog is heading towards double figures you should be anticipating his beginning to slow down, and making the necessary accommodations for his more elderly needs. As always, assess the dog in front of you, rather than making assumptions based on breed or type. You may have a 12-year-old Staffordshire Bull Terrier who is as fit and healthy as he was at 2, or a 7-year-old Lurcher who is grey-faced and noticeably slower than he was a couple of years ago. Age in human years is almost irrelevant: assess and adapt according to your individual dog's needs, and ensure that his remaining years are as good as they can possibly be.

It's important not to underestimate or ignore the emotional impact that caring for an older dog can have on you as the owner. For some dogs, old age is not problematic: they do not require any special adaptations or treatment, and they die peacefully when it's their time. However, for others, depending on the nature of their individual needs, their care may become much more complex, or they may be susceptible to unpredictable health events that reduce their quality of life dramatically with little or no warning. Not being able to plan for their future and the adaptations you may need to make is stressful; we all feel better when we have a plan and can take control.

The upsides of the older dog – that beautiful grey face, the slower pace of life, the joy he finds in being with you, those brief flashes of his younger self when you produce his favourite toy or take him to a favourite place – are numerous. Take the time to celebrate him as he was and as he is now, and discover the good moments you can still share.

I encourage you to find a vet you trust, and make the most of their expertise and advice for your older dog. Through your vet you can access many life-enhancing treatments and therapies, and make the best choices for you and your older dog. We are all more susceptible to illness as we age, and old age

Diva is the epitome of old-age beauty, enjoying her raised bed.

brings its own challenges: having a vet you can turn to for advice for your beloved oldie is priceless.

How to use this book

The book is designed for you to work through each chapter to get ideas and tips for living with and enjoying your older dog. You can, of course, dip in and out of the most relevant parts, but you will find a Quality of Life Scale in Appendix 2, that I encourage you to complete at the start, even if you have no current concerns for your older dog. Having this baseline can be particularly helpful as you approach making the decision of when to say goodbye, because it gives you a comparison for quality of life and the changes that have occurred.

    For ease of reading, please note that the main text refers to the dog as male throughout, although female is also implied at all times.

Resources
Books
Detector Dog: A Talking Dogs Scentwork Manual
 by Pam Mackinnon

Websites
Canine Arthritis Management https://caninearthritis.co.uk/
Galen Myotherapy https://www.caninetherapy.co.uk/

Facebook groups
Nail Maintenance For Dogs https://www.facebook.com/groups/nail.maintenance.for.dogs
UK Tracking Dog Association https://www.facebook.com/groups/TheUKTrackingDogAssociation/

/ Emotional awareness

Maintaining their emotional wellbeing as our dogs age is key to them living a happy and fulfilled life. If we can keep them feeling happy and joyful through appropriate social interactions, engaging in mood-boosting activities that make the most of their natural canine talents, and minimizing pain or distress, then their quality of life will be maximised.

Our dogs are sentient beings, with a full range of feelings, emotions, behaviours, and likes and dislikes. They can feel pain – which may seem like an obvious statement, but it's not long since this was first acknowledged – and have a right to express their opinions, and to be given choices and control in their lives. They are, however, not humans in fur coats, and it's important we ensure that what we offer them is appropriate for their needs, and we don't misinterpret their actions based on our human view of the world.

There is evidence to suggest their range of emotions is not as wide as ours, but that they share (along with all mammals) a large number of core emotional systems that enable and drive emotions. (For in-depth reading see The Emotional Foundations of Personality: A Neurobiological and Evolutionary Approach, Jaak Panksepp and Kenneth L Davis).

These systems have very specific scientific definitions, but for simplicity I have outlined them below –

- Seeking (expectancy): motivation to engage, move, search, eat, triggers anticipation, excitement and pleasure
- Play: joyful social interactions with other dogs, people; other animals
- Rage: frustration or irritation when the seeking system is obstructed
- Fear: closely linked to rage (fight or flight): purpose is to protect the individual from pain and destruction by avoidance
- Panic (grief): drives the desire to be close to owners or caregivers; linked with loss of security, loneliness
- Lust: search for sexual and social partners

Cooper enjoys the challenge of puzzle-feeders, like this Kong Gyro.™ It helps maintain mental flexibility as well as activating his seeking system.

- Care: parental care/nurturing of offspring

Optimum wellbeing for any dog or animal (and ourselves) involves maximising the positive systems and minimising triggering of the negative ones. For older dogs, the ones we focus on are maximising seeking and play and avoiding panic and fear.

If our older dogs are fearful, or feeling unsafe or lonely,

their quality of life is already compromised, and engaging in seeking or play is much less likely to happen, so we focus first on ensuring that their need for safety is met.

The most common sources of safety for our dogs are ourselves and any other dogs in the household. Dogs develop attachments very similar to the way in which humans do, and a secure attachment will provide them with a stable base from which to explore and enjoy the things life has to offer. At times, this secure attachment may become strained: either because they are too dependent on us (hyperattached), or because we can no longer provide the level of consistency they need.

Older dogs are particularly at risk of developing panic and fear-related issues because of the physical changes that accompany growing old. They are more likely to be left at home when the rest of the family (human and canine) goes for a long walk or day out that they can no longer manage. This may lead to increased separation-related distress where they have previously been happy to be left alone. They also have an increased likelihood of developing noise fears, which are often linked to being in pain, and so may be increasingly fearful, especially when you aren't there to comfort them. If their primary sources of safety (you and any other dogs in the household) are not available, we need to ensure they have an alternative: the safe haven.

As its name suggests, a safe haven is a place to go to feel safe. The motivation for using a safe haven is to move towards that place of safety. It is not a bolthole, where the motivation is to escape from whatever is scary or threatening. If your dog comes to you for comfort when he is scared, he is seeking you out as a safe haven because he will feel better with your attention and protection. If he hides under the table shaking at loud noises, but doesn't use that space at other times to rest and relax, it's more likely that under the table is a bolthole: somewhere to escape the noises but which doesn't make him feel better.

Creating a safe haven for your dog gives him comfort when you are not around, and can be portable, so that if you go to new places the safe haven comes, too. It will be your dog's favourite place to sleep and relax, and should be accessible whenever he needs it, but not in the busiest part of the house. Give him a space he can retreat to if needs be, but still feel part of the family. Avoid noisy areas such as alongside the washing machine/tumble dryer. A corner of the living room, behind or alongside the sofa, or under the stairs may be ideal.

Your older dog may prefer a slightly raised bed that supports him off the floor and out of draughts, or a padded, bolster-type bed that gives support for their head but is easy to step in and out of. Avoid loose blankets that he may trip over.

The companionship of other dogs in the family can be a valuable support for your older dog, providing safety and the sharing of positive experiences.

If you are using a crate, set it up so that, when open, the door won't catch on anyone or a dog walking past. Some dogs prefer the crate to be covered with a blanket, too. You may want to plug in a calming diffuser nearby – Pet Remedy or Adaptil, say – to help induce feelings of calmness and relaxation.

Once the safe haven area is set up, begin to introduce your dog to all the good things that happen there. Give him his meals in the bed; if you are going to give him a chew/treat, make sure it happens in the bed (you can hide these in there for him to find if you have only one dog). If he enjoys being massaged or groomed, indulge him in a pamper session in the bed; if he shows signs of being sleepy, encourage him into the bed, give him a couple of treats and let him sleep peacefully in his safe haven.

When your dog voluntarily takes himself off to his safe haven to sleep or relax, you'll know that he is beginning to like his new space. Reinforce this by dropping in a treat or a chew as you go past, but don't disturb him if he's sleeping. The more often he can sleep and relax there the better; he will develop a conditioned response to just being in that space where he will automatically feel relaxed and sleepy.

Do not tell your dog to go to his safe haven, or shut him in there, as any kind of punishment. If you need him to stay there for a while, give him a Kong™ or similar to keep him occupied, and encourage him to rest. While I love having my dogs sleep alongside me on the sofa when I'm relaxing, for their own

Older dog? No worries!

Pippin takes herself off to her safe haven when she needs a break or wants to rest. One side of the bed is lower to enable her to get in and out easily, and has supportive sides to cushion her back and head when she wants to stretch out.

emotional wellbeing, at times I want them to choose to sleep in their safe haven spaces instead.

Other considerations
As your dog ages he may start to bark more at what appears to be nothing, and be less able to stop himself doing so. He may also begin to withdraw from interactions with you, or with other dogs, or become less tolerant of boisterous interactions around him. He may start to toilet in the house because he was too slow to get outside, or because he is confused about where he should go.

With behaviours like these, or anything else you notice that's a new behaviour, book an appointment to discuss them with your vet. The changes could signal an increase in pain, preliminary changes related to Canine Cognitive Dysfunction, or other threats to his wellbeing which will need treatment. At all times these changes need our understanding; not punishment. Using any form of spray collar to punish barking, or raised voices to 'correct' apparently grumpy behaviour or housesoiling will be counterproductive, and introduce further fear and anxiety into an already confused or painful situation.

Maximising seeking and play
Alongside reducing fear and panic, we want to maximise all the good things in life via the seeking and play systems, maintaining positive interactions with the whole family.

Their safe haven doesn't need to be a crate. Riley uses her night-time bed during the day, too. To maximise her comfort, her bed rests on a folded duvet, and she likes to rest her head on the side for support.

Every bowl of food or puzzle-feeder, every walk, every toy, every interaction, activates the seeking system: however, some activities are particularly good at boosting mood, are highly reinforcing, and great for oldies. Of these, anything scent- or nose-related comes top for the combination of emotional wellbeing, mental stimulation, and being easily adaptable for any physical limitations.

Your dog is always using his nose, and it's often said that dogs 'see' through their noses, with complex scent patterns giving them information much as we would use our eyes to map an area. By tapping into his preferred method of information-gathering, and giving him plenty of opportunities to sniff out

Cooper and Pippin share a meal, sniffing out their dinner in the garden. Scatter dry food (or use small tubs if you feed wet food) around the lawn for them to find. Enjoyable activity plus physical exercise in one.

OLDER DOG? NO WORRIES!

things, we help to keep him feeling positive, and also keep his mental processes working as he ages. The act of sniffing and scenting is reinforcing in itself, and combining it with food or toy rewards, and teamwork with you, makes it even better.

Watching a dog searching, the movements of his nose, the changes in body language when he gets a sniff of the important scent, and the sheer enjoyment he experiences whilst searching is also hugely reinforcing for us humans, too. Anecdotally, scent-based activities appear to reduce incidents and duration of the fixed pattern pacing behaviour associated with Canine Cognitive Dysfunction.

Simple ideas to introduce more scent-based activities include –

- Scatter feeding their meals inside or outside, or using a snuffle mat, so they use their noses to find their dinner.

- Laying a treasure hunt, which can include finding his dinner, treats, chews, or a combination of all three

You can either lay out specific obstacles, different surfaces, textures, smells, and 'treasure' (food, toys, or a combination) and let your dog investigate on- or off-lead (depending on the area you're using and if it's safe for him to be off-lead), or use naturally-occurring obstacles and surfaces for your treasure area. If on the lead, let your dog guide you; don't force him to follow a particular pattern or route. If using toys, make sure you have a game with these when he discovers the toy.

- Tracking or trailing a human scent, with or without food or articles on the track/trail

Connected to you via a harness and a long line, let your dog follow a path previously laid by a person minutes or hours earlier. The track or trail can be as short and easy, or as long and complex as suits your dog, and can be done anywhere.

- Scentwork, sniffing for objects that have been infused with a specific scent such as catnip, or sniffing out pieces of cheese for non-toy-motivated dogs

There are multiple variants of scentwork, but my favourite, because it's specifically aimed at pet dogs, uses scents that are safe if accidentally ingested, and, above all, is fun and perfect for oldies, is Talking Dogs Scentwork.®

See Appendix 1 for more information on how to introduce scent and nosework to your life with your dog.

Cooper enjoys a snuffle mat, where his food has been hidden among the pieces of fleece knotted into a door mat. A great way to feed his dinner, or a snack, combining sniffing, eating, and problem-solving in one.

You can use anything you have to hand to create a new environment for a treasure hunt. Keep any obstacles low and stable, so that your dog won't injure himself if he stands on something that moves unexpectedly.

Play maintains mental and physical flexibility and strength, helps boost mood, and occupies time in what may otherwise be a dull day. When we think of play between dogs, most of us have an image of two or more dogs bowing, chasing, possibly playing bitey-face, or tugging at the same toy, often at high speed or for prolonged periods

When our dogs are young, this may be a regular feature of their social interactions with other dogs, but, as they mature, the frequency and duration of this kind of play reduces. Instead, the most common social interactions involve simply sharing time together – sniffing, walking; being together in close proximity but not necessarily touching.

Relationships between canine housemates can deteriorate as one dog ages, especially if younger animals are pushing past them through doorways or in other situations where excitement is high. An older dog can feel vulnerable, or may be in pain, which is aggravated by a younger dog forcing their way through: he may growl or snap in an attempt to create distance between himself and the younger dog.

In line with minimising fear and boosting play, we need to be aware of potential flashpoints, and teach some new house rules, for example, younger dogs should wait until the older dog is safely outside.

At the same time, encourage appropriate interactions

Older dog? No worries!

Try to keep up with old canine friends, and make new ones if your oldie is sociable and enjoys the company of other dogs. He may not play as he used to, but he will benefit from time spent just sniffing and mooching around with other like-minded dogs.

such as mooching around a field together, sharing a great sniff on a walk, or having a more relaxed game of bitey-face whilst laying down together.

To maintain positive play with people, you may need to modify some of your favourite games to make them appropriate for his current physical capabilities.

Repeated ball throwing, high-energy tug, lots of jumping on and off things put too much strain on older joints and will exacerbate arthritis or other painful conditions, so should be avoided. You may occasionally be tempted to do a couple of gentle throws of the ball, but leave the launcher at home. Likewise, if your dog adores playing tug, keep the sessions short and low-energy, and ensure that the tug stays in line with your dog's back ... and let him win. You can also use your play sessions to teach your old dog new tricks, using reward-based training methods, such as filing his own nails on a scratchboard, or

learning to use a dog ramp to ease his way in to the car or on to the sofa.

Your dog can also play solo with a range of puzzle-feeders and food toys, or with a specially put together box of delights. Find a box or bowl that is lower than his shoulder height and add a variety of different articles – toys, scrunched-up newspaper, empty plastic bottles, a stuffed Kong™, strips of fleece – anything to make the box interesting to your dog. Scatter some of his dinner, or treats, in the box, and mix things up so that he has to work to remove or move items to get at the food rewards. Vary the objects each time you give him the box to help keep it fun and fresh.

If using food toys and puzzle-feeders, you may notice that your dog is not as fast or adept at emptying these as he was. He may become locked into patterns that are not as successful, and may no longer have the mental flexibility – or the physical

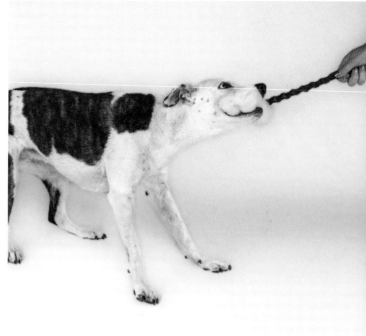

Ditch the ball thrower and revise any games of tug to reflect your older dog's physical limitations. Keep games for special occasions, and stay well within her limits. Keep the tug parallel with her back, and don't tug hard against her. Let her win the toy, and end the game sooner rather than later.

◀◀ *A Box of Delights can contain a range of objects, including food toys, toys for destroying, scattered treats and chews, strips of fleece, scrunched-up newspaper, empty plastic bottles: anything your dog can investigate and play with or remove to get to the things she really wants!*

capabilities – to switch tactics. If this happens, help him get the food, and next time you use that toy make it easier for him to do this himself, or swap to a different toy.

Pain

We cannot consider emotional wellbeing without also being aware of the impact that pain can have on how positive our dog may be feeling. Chronic pain and negative emotions often exist in a circular relationship, each making the other worse, and acute pain can have a dramatic impact on the seeking system, reducing all motivation to do anything. When depressed or anxious, our dog is more likely to feel pain than if he is feeling brighter or happier.

Slowing down, becoming stiff, showing less interest in exercise is not just part of getting older. Signs that your dog may be in pain include a reluctance to change position – to lie down if standing and vice versa – reluctance to go for a walk, or to play,

staying away from other dogs or you when relaxing, groaning when moving, an inability to hold a toileting position without moving, getting 'stuck' in corners/rooms, avoiding touch or fuss, tight facial muscles (the pain face), reluctance to eat or drink, and anything else that isn't 'normal' for your dog. If you notice any of these things, consider whether your dog could be in pain, and check him over from teeth to toes and tail together with your vet. There are multiple, new, targeted pain medications and specialist Pain Clinics at veterinary referral centres that can offer relief. Supplements may help reduce the stiffness or slowness, and there are many different therapies available, too, but please discuss with your vet, and be your older dog's advocate for a pain-free old age.

A word about routine and familiarity

The reduction in mental flexibility as they age may make an older dog very dependent on a routine or familiar environment, causing him to become anxious or distressed if his routine is disturbed. This situation is something you may have to maintain as much as you can, because he may no longer be able to adapt as he once did. For other oldies, trips to new places or a change of routine are just as fun or exciting as they always were, but don't forget to take along his safe haven, just in case.

Just as with us, as our dogs age, certain ailments and illnesses are more likely to develop, such as arthritis, tooth decay and gum disease, and obesity. To maintain their physical health for as long as possible there are several simple changes we can make to daily routines and our living spaces.

Husbandry care

As discussed in chapter 1, *Emotional awareness*, eliminating pain as much as possible is key to good wellbeing. Physical ailments are obviously the root cause of pain, so being aware of how your dog moves, his comfortable sleeping positions and general health is important. Your older dog may need more husbandry care to keep teeth and claws at their best, so include a daily check in your routine.

Tooth pain can be debilitating, and often shows as loss of appetite, reluctance to move, and avoidance of touch or contact with you. Your vet can advise and undertake a proper scale and polish, and may need to remove some already diseased teeth. Teeth should be white with little or no tartar, and gums should be pink with no inflammation. Check for bad breath, too: an obvious sign of gum or tooth problems. Ultrasonic tooth brushes are becoming very popular as an easy way to keep teeth free of plaque, or use a soft toothbrush and dog toothpaste to keep teeth and gums healthy. You can add seaweed-based supplements to food to help soften tartar, and there are also several dental chews available, although be aware of the additional calories you're adding to his diet if you use these, as being overweight is a risk to his health and comfort, too.

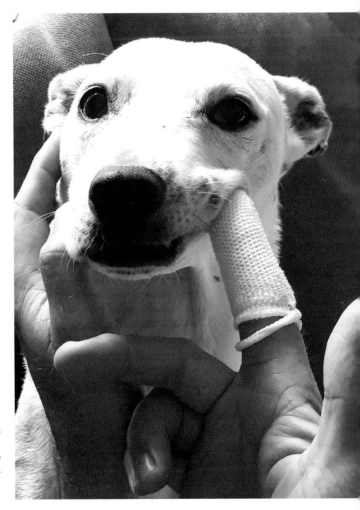

Make toothbrushing part of your regular routine with your older dog. Initially, just let him lick the toothpaste from your finger brush or toothbrush, and gradually progress to gently lifting his lips and rubbing the gums and teeth. Do not force him; take it slowly, and let him move away if necessary.

To teach your dog to enjoy nail trims, start by feeding her as you hold the clippers, and gently touch her feet and claws with them. Take the clippers away and stop feeding. When you see your dog begin to anticipate good things when she sees the clippers, you can progress to the next stage of gently holding a paw whilst feeding, and then onto using the clippers to cut a tiny sliver of nail. Cutting slivers avoids painful crushing of the nail, and equally painful cutting into the nail quick. Follow up each cut with a food reward: your dog will soon love having her nails done.

When he was younger and enjoying longer walks, it's possible you didn't need to clip your dog's claws, as pavements and tracks kept them naturally short. Short claws mean he can place his feet correctly, instead of trying to avoid the pressure of his claws hitting the floor and pushing back into his toes. Any adjustments he makes to his gait because of this has the potential to cause referred problems with joints and muscles, leading to pain and inflammation elsewhere or exacerbating existing joint issues.

Claw maintenance shouldn't be stressful for you or your dog. If he is already scared or reluctant to let you handle his feet, the simplest way to keep claws at the desired length is to teach him to use a scratchboard. You can make your own using a solid piece of wood, some medium sandpaper, and masking tape. You

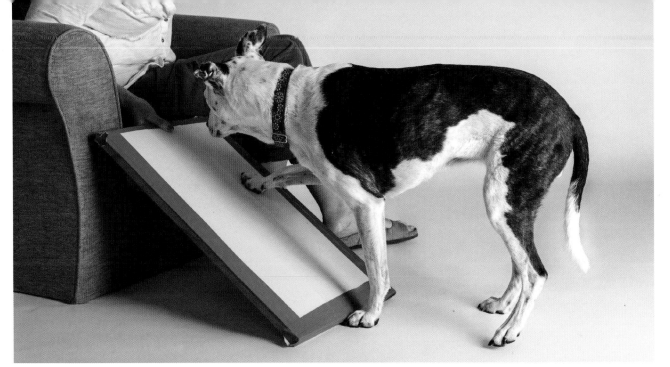

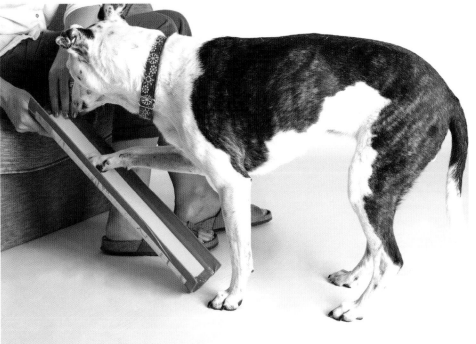

Keeping front feet nails short via a scratchboard or scratchpipe also provides mental stimulation as she learns how to use one. Start with 80 grit sandpaper while your dog is learning, and aim for 30-40 grit as she becomes more confident. A flat, wide board is easier for your dog when she is learning this technique as it's a bigger target for her.

Using a curved board, as shown in the picture left, helps to keep her outside nails as short as the inside ones. Alter the angle of the board as she scratches so that all nails are filed.

may also find a piece of guttering makes a good basis for a board because its curved shape files middle and outside toes evenly.

Hold a treat in your closed hand and place it on the scratchboard. As soon as your dog uses his paw to try to get the treat, open your hand and let him have the food. As he gains confidence with using his paw, hold your closed hand at the top of the board without a treat in it, and reward for scratching the board from the other hand. He will tend to swap legs naturally, or you can position the board so that he can only access it with the other paw. To trim hind feet nails, when your dog is relaxed and lying on his side, have a helper feed him favourite treats quickly and generously whilst you trim the nails. Take it one nail at a time; feed generously while you trim, then stop while you put down the foot and adjust for the best positon for the next nail. If necessary, do just one nail per session to avoid distressing your dog by trying to do more nails than he is comfortable with.

For more information about teaching dogs to relax and enjoy nail trims, see the Facebook group Nail Maintenance for Dogs.

Watching his weight

Your older dog is at risk of weight gain and obesity as his activity levels reduce and his metabolism slows down, but his food allowance stays the same. Being overweight is detrimental to joints, ease of movement; puts more strain on the heart, and has a negative impact on wellbeing. Similarly, being too skinny isn't healthy, either. If your older dog gains or loses weight rapidly, without any changes to activity levels or routine, discuss this with your vet: it can be a sign of health issues that require attention.

Take a good look at your older dog. Can you see a waist, and feel his ribs with minimal pressure? If so, it's likely he's a healthy weight. If you can *see* his ribs (some sighthound breeds

Lickimats and puzzle-feeders encourage your dog to spend longer eating his food, which can help if his portions are a lot smaller than they were because of his decreased mobility. Many dogs also prefer to eat lying down, as it reduces strain on their back and neck.

Older dog? No worries!

Using a raised feeder for water, or when bowl-feeding, will help alleviate any back or neck issues, and ensure your oldie can always comfortably access what she needs.

excepted), he is probably underweight. If you can't feel his ribs, or he has a pot belly, he is overweight.

Remember that feeding instructions on dog food are only a guideline, and weighing every meal is the only accurate way to ensure you stick to the correct amount. Use the manufacturer's recommendations as a starting point, but adjust up or down as necessary for your dog. You may end up feeding two dogs of the same weight very different amounts of food to keep both at their optimum weight. If you have given a lot of treats, or additional chews, reduce the amount of food given to compensate. This is another reason for using puzzle-feeders or a Lickimat instead of a bowl to feed: even a small portion requires effort, and takes longer to eat instead of being gone in two gulps.

While assessing your dog's dietary needs, you may also consider adding a joint supplement such as Riaflex or YuMove to help support his mobility.

Adaptations to your home and car

Your oldie may have claws of the perfect length, but can still be unsteady on his feet, and find smooth flooring such as laminate, tiles, wood, and some vinyl very slippery underfoot. This lack of traction makes it harder for him to stand, sit, lie down or walk around, and increases the risk of injury or additional soreness from slipping.

The easiest solution, to avoid the need to replace all your hard floors with carpet, is to use large rugs to cover walkways and ease your dog's movements around the home. Ensure the rugs are rubber-backed, or use special anti-slip underlay so they don't move when walked on.

If your dog is used to sleeping on the sofa or your bed, he is also going to need help to access favourite spots, as well as a way of getting in and out of the car safely. There are lots of options for dog ramps, but make sure that whatever you get is stable, wide enough for your dog to walk on comfortably, and non-slip.

Start with the ramp laid out flat on the floor, and place some treats on it for your dog to investigate. When he is quite happily placing front and back feet on the ramp, use the treats to encourage him to walk along it whilst it is still on the floor. Gradually increase the slope of the ramp – try resting on sofa cushions at first – until your dog is comfortably walking up and down at the height you need. Always stand alongside him, and be ready to support as needed.

Depending on your dog's particular mobility issues, he may prefer steps to access the car or the sofa.

There may come a time when your dog can no longer comfortably climb the stairs to your bedroom. You may consider moving your sleeping arrangements downstairs so that you can still be together, or can you safely carry your dog up and down, potentially several times a night if he needs toilet breaks, without hurting or stressing him? Harnesses are available that can help support, but you will need to consider how taking it on and off will impact on its usefulness, especially if your dog needs to go outside to toilet urgently.

You may decide that it is safer for your older dog to sleep downstairs in his safe haven. In order to make this transition easier for everyone, plan to spend the first few nights downstairs with him. Give him a small chew to settle him in his safe haven, and sleep alongside him initially. Once he has settled and is sleeping through for a couple of nights, you can go back to sleeping upstairs.

How does your dog access the garden for toileting? Is there a step outside the back door? Is the surface outside easy for him to walk on, or are there uneven areas or obstacles to negotiate? Older dogs are likely to need to toilet more frequently, so the easier you can make this the better. Add a ramp instead of a step, move any obstacles, use some matting or astroturf temporarily to create a more even surface.

You may also notice that your dog no longer stays in one place while toileting; this is an indication of pain and/or muscle weakness, which means he can no longer hold the necessary position for long enough to eliminate fully. Speak to your vet for possible support for your dog to reduce his pain.

Tess is comfortable using the ramp, fully supported by her owner to help prevent slipping or stumbling. The ramp enables Tess to continue enjoying walks and trips out in the car.

Older dog? No worries!

Riley prefers to use steps to get in and out of the van. The steps are non-slip, and fold up for ease of transport.

Make it as easy as possible for your oldie to access the garden for toileting: use ramps or shallow steps covered with non-slip matting to provide even surfaces and stability.

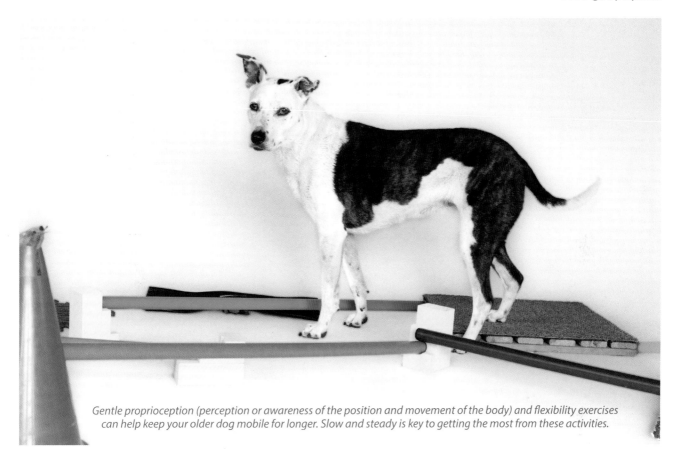

Gentle proprioception (perception or awareness of the position and movement of the body) and flexibility exercises can help keep your older dog mobile for longer. Slow and steady is key to getting the most from these activities.

Keep on moving

As with everything, prevention is better than cure, and ensuring that your dog is a healthy weight, is fit and appropriately active will help prevent problems in the future. If your dog is aging but has no existing health or mobility issues, begin to introduce simple proprioception (perception or awareness of the position and movement of the body) and balance skills to your walks and games time.

Use a variety of surfaces, textures, cones, poles and movements such as around to the left/right, under or over, two paws on/four paws on, bowing, side-stepping and reversing to gently work each muscle group and maintain muscle condition. See how slowly your dog can do the various movements – this is not a speed exercise. Move slowly from down to stand and back

again (use a treat to lure as necessary) keeping all four paws in the same position. Lure a slow sit to stand, and back again.

Remember your dog is using his muscles, not momentum, to do these exercises slowly, so don't do more than one or two repetitions, and let him take a break whenever he wants to. Pay attention to what he is telling you: if he is reluctant to move in a particular way, it may be because it is painful to do so. If you have any concerns over his movement, speak to your vet.

Your older dog will inevitably slow down on walks, be unable to walk as far or as fast as he once could, and reduce the amount of play he is willing or able to engage in with you or other dogs. As discussed in the *Emotional awareness* chapter, you will need to alter play and game time to suit your older

continued page 28

◀◀ *Bacon has done these kinds of conditioning exercises all her life, and is definitely an expert. Start with a shallow step, lower than the height of your dog's stopper pad, and only increase the height and difficulty if your dog has no physical problems. Never use inflatable equipment unless instructed by a canine physiotherapist: this is not necessary, and can cause injuries if used incorrectly.*

The Senior Social Club members enjoy their regular get-togethers in a safe, enclosed field. There's very little running around, but lots of mutual sniffing, pottering, pond-dipping, and asking for treats from all the humans. A great trip out!

dog, and similarly your walks will change. Your older dog will prefer shorter walks carried out more frequently during the day, instead of one longer walk. Twenty to thirty minutes at a time of slow sniffing and pottering, with opportunities to interact with familiar dogs and people if he enjoys that, is an ideal walk for many oldies. Replacing a walk with a short car journey, with the opportunity to watch the world go by at your destination, can also be very beneficial and enjoyable, because it gives the opportunity for new smells, mental stimulation, and emotional enrichment without excessive physical impact.

Canine physiotherapists and hydrotherapists work on vet referral, and provide an invaluable source of help and support for your older dog. It is hugely beneficial to introduce your dog to physiotherapy and hydrotherapy support while he is relatively fit and healthy, so that he is relaxed and comfortable receiving treatment before it is actually needed. Trying to teach a dog who is in pain and uninterested in food or toys that a hydrotherapy treadmill is a safe and enjoyable place to be is a huge challenge. Introduce the hydrotherapy equipment – many therapists will offer special introductory sessions where the emphasis is on fun and relaxation rather than treatment – before your dog actively needs help.

Your physiotherapist will fully assess your dog's needs, treat appropriately, and give you a programme of exercises if necessary that will help improve your dog's physical condition and quality of life.

For general mobility and comfort, treat your dog to regular massages from an appropriately qualified massage therapist, or learn to do this yourself: it will be an enjoyable experience for you both.

Lifting and aids to movement

When mobility becomes a more serious problem, there are several options to help prolong your dog's ability to move independently, or with support from you.

A physiotherapist can show you how to safely lift your dog, depending on his size and physical problems, and help you ensure that he is comfortable being carried.

Dog wheelchairs are now available in a range of sizes, with options for rear-end only or front and back leg support, and

Introduce your dog to a hydrotherapy treadmill before she needs treatment, so that if she is in pain or injured, the treadmill is a familiar and comfortable environment rather than something that adds to her stress. With thanks to Fentem Jones Veterinary Rehabilitation for the photo, and for helping to keep Riley mobile and comfortable.

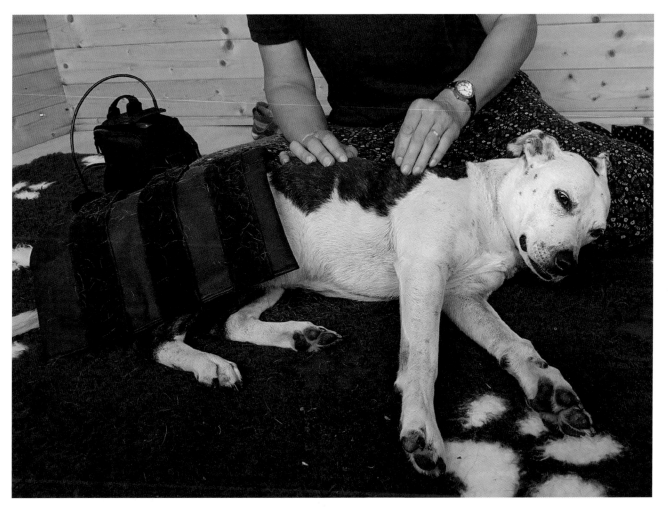

Massage therapy, including magnet and laser therapy, can be hugely beneficial for older dogs. As with all treatment, listen to and watch your dog and allow her to move away if she wants to. With thanks to Nicky Wreford Physiotherapist and Galen Myotherapist for the photo, and helping to keep Riley mobile and comfortable.

are well tolerated by many dogs. If considering wheels for your older dog, always bear in mind he will need your help to get in and out of these, and how well he might cope with this. Also consider how you will assist him with toileting or moving around when he is not wearing his wheels. Wheels can be very life-enhancing for some dogs, but they also limit their choices (your dog cannot choose to lie down and rest while in his wheels), so should be carefully assessed on an individual basis.

There's also a range of harnesses to help support front end, back end, or both, which can help you keep your older dog moving, helping him up stairs or ramps, or supporting him while toileting if he is unable to support his own body weight.

continued page 33

Older dog? No worries!

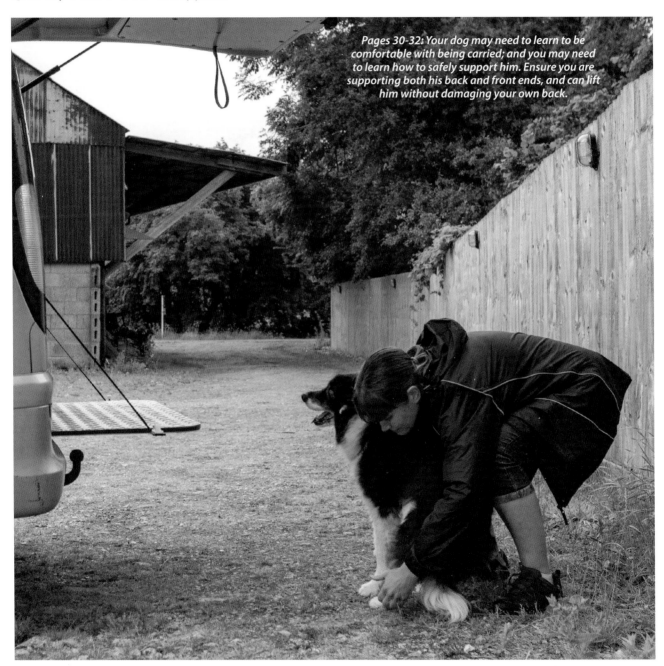

Pages 30-32: Your dog may need to learn to be comfortable with being carried; and you may need to learn how to safely support him. Ensure you are supporting both his back and front ends, and can lift him without damaging your own back.

Older dog? No worries!

Some older dogs also love their buggies/pushchairs, and these can be ideal to help your dog carry on enjoying walks with your family, and giving him opportunities to rest while everyone else carries on walking.

With all the options for mobility aids and support, take care to introduce these to your dog as early as possible, in a gentle and positive way that allows him to become familiar and comfortable with the equipment. Always listen to your dog and, if he is reluctant, uncomfortable or fearful, enlist the help of a reward-based trainer (a member of the APDT UK, for example) to help you help him to relax.

Think PAL
If you have any doubts or concerns about your older dog's mobility and physical capabilities, think PAL and act accordingly –

Pain relief
- Discuss with your vet
- Supplementation: eg Riaflex, YuMove
- Hydrotherapy/physiotherapy
- Laser/massage therapy

Adjust activity
- Shorter walks
- Mobility exercises
- Car trips
- Diet and feeding

Listen to your dog
- Unsteadiness or skipping (not using a back leg when moving)
- Reluctance to change position
- Getting 'stuck'

3 Stimulate the senses

Hearing, sight, taste, touch, and smell: if you're a dog, the greatest of these is smell. Your dog has the same senses as humans, and, like humans, their senses are likely to deteriorate with age. The changes to their ability to hear and see can be the most obvious, and have the greatest impact on their quality of life and engagement with the rest of the family. All of the senses can enhance wellbeing and maintain quality of life.

Hearing

You may notice changes in your dog's hearing when he is relatively young. As discussed in *Emotional awareness*, noise fears typically develop at around 4-5 years of age, and this may be because a dog's hearing changes as he ages, and certain pitches or tones of noise may become more disturbing. It can also be linked to pain experienced when a loud noise startles him and causes him to jump, contracting muscles or triggering pain in sensitive areas. The noise then predicts that he will feel pain, and he begins to anticipate and fear the noise happening again. If your dog is fearful of noises, speak to your vet about a physical check-up, and appropriate treatment for pain relief.

There are, of course, many dogs who are deaf from birth, who live happy and fulfilled lives: with a few adaptations from you, your older dog can do so, too. It is important to be aware that a deaf or hearing-impaired dog is more likely to startle if you approach or touch him when he hasn't seen you, and in some cases this may lead a dog to snap in surprise. Never stroke your deaf dog whilst he is resting or sleeping; leave him to relax without interference. Older dogs will sleep more deeply, even if their hearing isn't impaired, so keep your distance, and don't allow other dogs to disturb him either.

Reinforce your dog every time he orientates towards you, looks at your face or makes eye contact, using food and attention where possible. You want him to develop a habit of checking in with you so that it becomes second nature for him to look to you in all situations.

Riley can still enjoy paddling in the sea whilst wearing a long line. Biothane lines like this one are waterproof and lightweight, so don't become heavy when wet: ideal for the beach and wet grass. Bright colours are easily seen by other walkers so they know your dog is on a lead.

One of the biggest impacts that losing his hearing can have on your relationship with your dog is that he can no longer hear you when you come home, call his name, prepare his food, or try to get his attention. This can take some adjusting to as you will need to change how you position yourself in front of him, how you get his focus on you, and how you ensure he is safe whilst out and about.

As a starting point, this is a good time to refresh early training to encourage your dog to check in with you automatically, rather than only when you give a 'watch me' type cue. Carry some tasty treats, or part of his daily food allowance, with you at all times, and every time your dog looks at you (he doesn't have to make eye contact; just look in your general direction), give him a 'thumbs-up' sign and a treat. You should find that he quickly begins to pay attention to you a lot more often, because it has become a very worthwhile thing to do. Practice this in the house, garden, on walks, and everywhere you go so that he gets used to engaging with you in lots of different places.

As dogs lose their hearing, you can no longer rely on this to recall them on walks. The safest and easiest solution to this is to fit your dog with a suitable walking harness, such as Perfect Fit or Mekuti, and attach a long line to the back ring so that he can have a degree of freedom to potter around and sniff, but can't wander off, or accidentally go out of sight if something catches his interest. Continue to reward him for checking in with you on walks, even when on a long line.

Use visual instead of verbal cues (see *Retain the brain* for more ideas) for any behaviours that you want to teach or maintain, and carry on talking to your dog as you always have done. He may or may not be able to hear you, but the chances are he will feel the vibrations (especially if you are touching or stroking him at the time), and will see your face and mouth moving as he always has done. Dogs are masters of observation, and it's likely that your facial expressions have always told him a lot more than the noises coming out of your mouth, even when he had full hearing.

A family walk in the woods using a long line so that Riley can stop and sniff when she wants to, and has some freedom, but can't become lost if she wanders too far out of hearing and sight.

Older dog? No worries!

There is evidence to suggest that music helps to relax and calm dogs of all ages, reducing heart rate and encouraging resting behaviour: Spotify even has its own dog music channel, and YouTube has plenty of options to try out. Use music when he is home alone, or to encourage rest at times when he may be pacing (a common symptom of canine cognitive dysfunction), or as part of a night-time sleeping routine. As with your voice, it is likely that hearing-impaired dogs will feel the vibrations, even if they can't hear all of the notes.

Sight

As with hearing, a dog's eyesight is likely to change and gradually decline as he ages. The most commonly noticeable change is apparent cloudiness of the lens of the eyes that develops from around the age of 7 onwards. This is usually normal thickening of the lens that occurs in all dogs, but can also be the development of a cataract: an abnormality that should be treated.

Normal vision changes in old age reduce your dog's

▸▸ Always check whether your dog wants you to continue fussing her. Stop touching her and wait to see if she asks you to continue – Riley uses her paw to prod us to carry on. If she doesn't ask for further touch, leave her to rest in peace.

ability to see close-up, just like humans needing reading glasses, and may impact on his ability to catch a treat, or navigate stairs. It may also change the way he looks at you, especially if he is sitting with you on the sofa and your face is close to his, because he can no longer focus close-up.

Night-time vision is also likely to deteriorate, so he may be unsettled at dusk and less willing to go outside then/ in the dark, or hesitant in new areas. Adding additional lighting or accompanying him outside for reassurance will help. You may also notice that the pupil (the dark central part of the eye) appears larger as the muscle that controls its contraction deteriorates with age. If you have any concerns over the cloudiness or appearance of your dog's eyes, discuss this with your vet.

Most older dogs are still able to see well enough to recognise visual cues from people, and to move with confidence around familiar and unfamiliar spaces. They can also see yellow and blue much more distinctly than other colours (green and red appear as shades of grey), so choose toys or mark edges and steps in these colours to help them stand out.

If your dog does lose his sight, you can help him by maintaining the familiar layout of his home, and using verbal cues (as long as his hearing is still reliable). If both senses are failing then touch-based cues where appropriate (when your dog is not sleeping or focused away from you), and scent-based cues can help maintain quality of communication between you.

See *Retain the brain* chapter for how to introduce new cues for existing behaviours to ensure you maximise your dog's remaining senses.

Touch

If your dog's hearing and sight have declined, his ability to feel you stroke him, or to sense your proximity using his whiskers and vibrations, is unlikely to change, and may even improve to compensate for the lack of other senses. Touch can become increasingly important to both of you, which is a further reason

Providing a nightlight can help your older dog settle in the evenings/at bedtime. Many older dogs begin to pace in the evenings, due to Canine Cognitive Dysfunction, and having a light in their sleeping area can reduce the pacing and encourage rest.

to ensure that any pain is under control so that your dog can enjoy physical interactions.

Simple massage techniques, such as long, slow strokes down his body, with one hand maintaining contact with his fur at all times, can promote feelings of relaxation and wellbeing in both of you. For more active times, a simple nose-to-hand touch helps guide your dog along the easiest route for him. To start to teach him to place his nose on your hand, hold a treat between the outstretched fingers of a flat hand, and offer it to him to sniff. As his nose touches your palm, open your fingers slightly to release the treat. Repeat a few times, and then offer your hand in the same way but without the treat. As his nose touches your hand, feed a treat with your other hand. Gradually increase the distance between your dog and your hand so that he has to walk a short distance to make contact with your hand. If necessary, you can continue to lure him, so you always have a treat between your fingers, while he keeps his nose in contact with your hand. You can use this technique to move him without taking hold of his collar or harness; it can be particularly useful if he is anxious, or reluctant to move.

Taste

Changes in appetite in older dogs can be one of the most stressful aspects of their aging; feeding them is one of the most basic elements of our caring relationship with our dogs, and watching them refuse food, or having to reject their repeated requests for extra food, is upsetting. Any dramatic changes in appetite, such as increased hunger or food refusal, should be investigated by your vet, as should any changes in weight without changes to the amount of food he is receiving. Likewise, your older dog should drink around the same amount he always has done (assuming no changes to diet), and any differences should be discussed with your vet.

Once you have ruled out physical reasons for changes in appetite: for example gum/tooth infections or neck/back problems that make eating uncomfortable, you may still need to tempt your dog to eat at times. Things to try include adding bone broth to dry food, swapping to a wet food, or mixing together wet and dry food. You may also want to swap to a senior diet dog food with slightly lower protein levels. These diets can be higher in fat – to increase their appeal to fussy eaters – but are also often developed to help dogs feel fuller for longer as feeding amounts are reduced to reflect their decreased activity levels. Finding the right food can be a case of trial and error, but be guided by what you see and feel of your dog's condition. If he is maintaining an appropriate weight, apparently enjoying his food, and not experiencing any upset stomachs or

Practice your nose touches everywhere, and always reward with a treat. You want this to be a very easy thing for your dog to do, and always associated with good things. Keep your hand at the same height as your dog's nose so she doesn't have to strain to touch.

constipation, then it's likely that you have found the right food for him.

There are multiple options for supplementation to optimise your dog's diet. Many older dog diets contain additional omega 3 oils and selenium to support brain function, glucosamine and chondroitin for joint health, and additional anti-oxidants for cell protection. Adding pro-biotics to boost gut health is also common.

Feeding via enrichment feeders, snuffle mats, scatter feeding, or treasure hunts can help fussy eaters maximise their intake. Freely available food often loses its appeal, precisely because it's too easy and always there, whereas something they have to work for, or that is more rarely available, increases in value. For the hungry dog, working for his dinner slows down his eating, provides additional mental challenges, and keeps him occupied. You can also bulk out the hungry dog's diet with chopped raw vegetables such as carrot, broccoli, and cauliflower to add volume without calories.

As her sense of taste changes, you may need to experiment with new treats and chews; sometimes the smellier the better. Diva is particularly fond of crunchy fish skins, which have the added advantage of cleaning teeth and gums, and providing omega 3, which is good for the brain.

Freddie enjoys using the Lickibowl for his meals. The wet food sticks to the sides, which means he needs to lick it clean, but the bowl moves around, making this more of a challenge.

Scent

Fortunately for most dogs, scent remains the primary sense throughout their lives, and while they may lose some of their abilities as they age, the nose generally remains their most informative and powerful organ. There is so much we can do to help our older dogs remain engaged, active and happy through harnessing the power of their nose. Appendix A describes in more detail how to introduce nose-based activities such as scentwork, tracking, and trailing, and some ideas to use scent in everyday lives include –

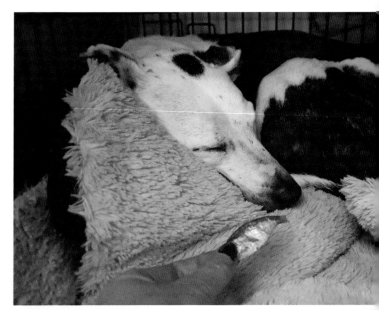

- If your dog is sleeping, wake him gently by holding a smelly piece of food a small distance away from his nose. This will generally rouse him more effectively and kindly than touching him or making a noise that may startle him

- You probably won't need to call him for dinner, as the scent of you preparing food will catch his attention, but if he is losing sight and hearing, wafting his food bowl in front of him will encourage him to move

- Substitute known verbal or visual cues with scent-based cues: this is excellent brainwork as well as making the most of his nose. See *Retain the brain* for how to do this with your dog

- Include plants in your garden, or in tubs, that may appeal to your dog: birch, catnip, lavender, marigold, meadowsweet, peppermint, valerian, wheatgrass, and willow, for example

- Bring new scents home with you. If your dog can't go on a family outing, ask people to hold a piece of cloth in their hands for a few minutes, then put each one in a separate bag for your dog to sniff when you get home. You can also rub cloths over other dogs, cats, horses, or any animal who doesn't mind being petted by you to take the scent back for your dog to enjoy

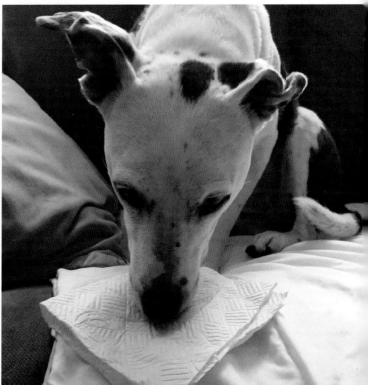

❤ Riley is fast asleep, but I need to wake her so she can toilet before I have to go out. Use something really smelly – in this case a dried sprat – which will waft into her nasal passages and wake her gently.

▸▸ A piece of kitchen towel that has been held by a person, or wiped over another dog/animal, gives your dog a new scent to take in and enjoy. If you go out without your oldie, try taking things back for her to sniff.

OLDER DOG? NO WORRIES!

Temperature regulation

Whilst not strictly one of the primary senses, their sensitivity to hot and cold changes as dogs age. In cold weather dogs are much more likely to need to wear a coat or jumper to stay warm, and, conversely, older dogs will struggle to stay cool in warm temperatures. Regulating body heat requires a lot of energy to be expended via shivering or panting, and is much harder for older dogs.

In winter, check for cold ears, paws and tummies, and pop a jumper on him, or turn up the heating as necessary. In the summer, use cool mats, cool coats, and shade, and make sure there is plenty of water available to help your oldie stay cool. If necessary, wet the coat around the back of the head and neck, or provide an easily accessible paddling pool so he can cool off when he wants to.

Your older dog may feel the cold more than he used to, and may need to wear a jumper even indoors, or overnight. If he wakes at night, try pyjamas to see if being warmer helps him sleep through.

4 Retain the brain

Use it or lose it is true of mental as well as physical capabilities, so the more activities and experiences you can do and have with your older dog, the better it will help retain his mental acuity. Most dogs will lose some of their mental function as they age; however, Canine Cognitive Dysfunction (or Doggy Dementia) is increasingly recognised as a common problem for aging dogs, and goes beyond normal old age memory loss. There is no cure for CCD, but you can help to delay its progression, and make living with a dog with CCD easier to cope with.

Use **DISHA** to help you assess if your dog is showing signs of developing CCD –

Disorientation – getting 'lost' or 'stuck' in familiar places

Interactions – irritability; withdrawal, lack of interest

Sleep changes – wakefulness in the night; unable to settle/rest

Housesoiling – being unable to hold on/unaware of what he is doing

Activity changes – increased barking, pacing; repetitive behaviours

Your dog may have one or more of the above signs, and it's also worth bearing in mind that pain can manifest as some or all of the above indications. As always, speak to your vet about assessment and treatment as successful medical interventions exist that can alleviate the symptoms of CCD.

The nose-to-hand touch described in *Stimulate the senses* can be one of the most useful tools for managing a dog with CCD; it can be used to help interrupt any of the repetitive barking or pacing behaviours by giving your dog something to focus on and engage with. It's also useful for helping to move him if he becomes 'stuck' or locked in to moving in a particular

pattern. You can move him towards his bed, or help get him out of the bed in the first place using a 'touch' cue. If you find your older dog develops a repetitive movement pattern – he always takes a particular route around the garden, repeating the track multiple times without stopping to sniff or interact with anything – it can be helpful to re-orientate him using the hand touch, and then give him something else to occupy him.

There may be times when you will need to lead him to his safe haven, and give him a chew or similar, with the door closed if necessary to limit his movement.

There is also anecdotal evidence that scent-based activities such as sniffing for hidden scented toys helps with CCD, and this may be because the scent games activate the seeking system, and provide an endorphin release that reduces the need for the pattern-making pacing.

New experiences

Assuming any pain is fully controlled and your older dog is engaged, confident and inquisitive, he can still enjoy going to new places and meeting new people. A favourite for our oldie is to go to work with my husband for a day every few weeks, where she can watch what's going on, sleep under a desk, be spoiled rotten by colleagues, and potter along the river bank investigating swan poo and new smells.

If yours enjoys a car ride, get him a crash-tested car harness and let him sit on the passenger seat so that he can see out, and be your co-pilot on a road trip to a dog-friendly café or farm shop where he can help you do your shopping, or just hang out together. Instead of a road trip, or a day out walking, how about a boat trip where you can both watch the world go by, have lunch at a pub or picnic along the way, and your oldie gets lots of mental stimulation for minimal physical effort.

You can still include your older dog in family holidays; make sure you take his safe haven and any familiar scents you use to help him navigate in your house, and pack the enzymatic

OLDER DOG? NO WORRIES!

Found it! Searching for catnip-scented toys is one of Riley's favourite things to do. She lights up when I pick up the scentwork tin, and it's a great focus of mental and physical enrichment for her, along with the emotional boost that comes from sniffing. She also paces less in the evenings on days when she has done some scentwork

Old ladies get front seat privileges, even on a boat trip. Watching the world go by with Dad on the Norfolk Broads.

All the sights, sounds, and scents from the back of the boat for Cooper. With a stop at a pub for lunch, this was a perfect day out for the oldies.

cleaner just in case of any accidents in the unfamiliar place. Always make sure the first thing you do in a new place is show him how to find the garden or toilet area, and lead him to his safe haven, with a chew or puzzle toy to help him settle in.

Old dog; new tricks

It's likely you taught your dog simple behaviours like sit, down, and stand at puppy class, or when he first came to live with you, so refresh his memory about these and any other known cues once or twice a week. Vary the order in which you ask for the behaviours – you'll be amazed at how quickly even older dogs learn to anticipate and offer what they think will be coming next – and remember that some physical behaviours may be harder for your dog to do, so if he is reluctant to lie down or slower standing up, make allowances for him.

Of course, it's possible to teach your older dog new things; even if he may not learn things quite as quickly as he once did, you can still have fun learning together. Combine mental stimulation with physical exercise to maintain his mobility in a dual purpose activity such as teaching clockwise and anti-clockwise spins that encourage flexibility in the neck and spine, work your dog's muscles evenly on both sides, and require some brain power to remember which is which.

Begin by luring your dog clockwise, holding a treat in one hand and moving it slowly in front of his nose in a clockwise direction. Reward from your hand when he moves to follow the treat, and ends up back where he started. Repeat, but go anti-clockwise the next time. Always do both sides when practising this exercise, even though your dog is likely to find one side easier than the other: it will help to maintain even muscle condition.

There are lots of ideas for mental stimulation and tricks to entertain your dog; the following are just a few suggestions which can be adapted to suit all levels of activity and engagement.

Yellow/Blue

Teach your dog to discriminate between yellow and blue (the two colours that are most different on the canine colour spectrum), and then to apply this to any yellow or blue object you show him.

Fix a yellow post-it note on the end of a wooden spoon, and show this to your dog, holding it at the same height as his nose, though do not touch his nose with the spoon. Let him look

Help to refresh your dog's memory by practising all of the tricks and cues she learned when she was younger. It doesn't matter if you go back to luring her to get her motivated: it's about having fun together, seeing what she can do, and maybe adding to her repertoire.

at or approach the post-it and mark this by saying 'good' as he does, then give him a treat from your other hand. Present the spoon and post-it again, and mark that he has done the right thing with 'good' and reward as before. Next, only say 'good' (mark) when his nose touches the post-it, and always follow up with the reward.

Make it easy, at first, for your dog to touch the post-it with his nose; gradually making it harder to do so by his having to move slightly, or turn his head in order to touch the post-it with his nose. When he is confidently hitting the post-it each time, begin to add the cue 'yellow,' saying 'yellow' and then presenting the post-it note on the spoon. Mark when he touches this with his nose, and then reward. Repeat five or six times in each training session and then give him a break.

When he is confidently looking for the post-it when you

say 'yellow,' start to place the note on other objects, such as on a cupboard door, a chair, or your foot. Once he is touching the post-it regardless of where it is, repeat the exercise above, but this time using a blue post-it note and, obviously, the cue 'blue.'

You can then teach your dog to generalise the yellow and blue cues to other yellow and blue objects. Place a yellow post-it note on a yellow object – a mug, cone, welly, say – and show it to your dog with the cue 'yellow.' Mark and reward as before when he touches the post-it. Repeat this a few times, then remove the post-it note and present the yellow object again with the cue 'yellow.' Mark and reward if he touches anywhere on the object.

If he seems confused, go back a step and do more repetitions of cue, present, mark, reward with the post-it note on the object. Try removing the note once more and giving the cue

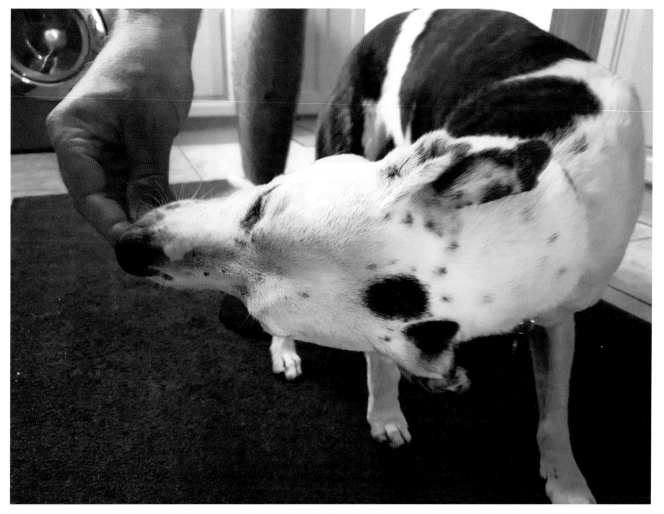

Spins and twists, done slowly and with precision, help increase flexibility in the spine, as well as providing a mental challenge in remembering which is clockwise and which is anti-clockwise.

'yellow' as you present the object again. Repeat the above steps with blue objects and your 'blue' cue once 'yellow' has become routine.

The next step in this training game is to ask your dog to touch the correct colour object when both options are in front of him: one yellow and one blue cone, for example. This can be the trickiest part for your dog to get right, so take it slowly. If he is making multiple errors you are moving too fast, and need to make it easier again.

Place your blue object on the floor so that your dog sees you do it, but at a reasonable distance from him; then place your yellow object on the floor so that it is much closer to you and your dog. Give your 'yellow' cue and mark (say 'good') any movement towards the yellow object, and reward. If your dog

OLDER DOG? NO WORRIES!

◀◀ Freddie has mastered the Yellow/Blue challenge, and correctly indicates the blue cone when presented with this as a cue.

is confidently touching the yellow object when you cue 'yellow,' move the blue object slightly closer every couple of repetitions so that he is learning to correctly choose between them. Begin to vary the positions of both objects if he continues to choose correctly, moving the yellow object further away to increase the difficulty.

If he touches the blue object rather than the yellow one, don't mark this in any way, but move the blue object further away before again cueing 'yellow.' You want him to learn with minimal errors so that he doesn't just guess.

Once your dog is confident with yellow, repeat the above routine with the blue object and your 'blue' cue. As he will have had lots of practice with the yellow object, he may be less willing/confident to touch the blue one: ensure that the yellow one is a long distance away at first.

Having successfully taught your dog to discriminate between colours and a variety of coloured objects, you can add a final element to this game whereby he learns to match not only the colour but the object or shape, too. Place a yellow cone and a blue cone on the floor, and hold up another blue cone. Give

your 'blue' cue and watch your dog touch the blue cone. Give yourselves a huge pat on the back and a party if you get to this stage; it's a great achievement for *any* dog.

FIND THE LADY

Place a treat underneath a cup in front of your dog, and move the cup around before encouraging your dog to get the treat. Do this a few times until he is confidently knocking over the cup to get the treat, and then add a second cup. Place a treat under one of the cups, swap the cup positions, and encourage your dog to find the treat. Did he get the right one? Try again, and add a third cup if two is too easy for him.

This is a game, and can be a great way of getting some dogs to eat their dinner, so it's not important if they get it right or wrong as long as you are both having fun, and they get it right often enough for the treat to motivate them to try again. Make it easier or go back to just one cup if they are struggling.

WHERE'S WALLY? (OR ANY OTHER TOY)

Knowing his favourite toys by name, and being able to indicate

This may just be a game, but it also illustrates that dogs have a concept of object permanence – things still exist even when they go out of sight. Riley chooses which mug is hiding the treat.

She has to work to knock over the mug to get the treat; sometimes they end up all over the kitchen floor!

When asked to, Riley correctly picks Toppl from among a selection of her toys.

or bring you the correct one from a pile on the floor or his toy box, is a fun trick that's easy to teach. Every time you interact with your dog and the toy, use the name you've given it (tyre, Toppl, squirrel, etc). Repeat, repeat, repeat this so that he makes the association with the sound that's coming out of your mouth and the toy you're holding or pointing at. Ask "Where's Toppl?" and set off to find it with him, encouraging him to pick it up or grab it when you find it. For dogs with reduced mobility you can play the same game with them sitting or lying down, and placing a row of toys in front of them. They just need to look at or nose towards the correct one before you pick it up and give it to them.

ADDING OR CHANGING CUES
Once you understand how dogs learn to identify a cue for a behaviour, you can use this simple technique to develop all sorts

of impressive behaviours and tricks with a range of different types of cue.

Whenever you teach your dog a new behaviour, you want to give it a cue. This could be a verbal or aural cue – a word or a whistle; a visual cue – a hand signal; a contextual cue – being at the gate to his favourite field; a scent cue – camomile, or a touch cue – your hand on his shoulder. It is best practice to only add a cue to a behaviour once your dog is performing the behaviour in exactly the way you want him to do it: when teaching 'sit' we use a hand signal, initially, until the dog really understands he should place his bottom on the floor when asked to, and then we swap the hand signal for the verbal 'sit' cue that we really want to use going forward.

All dogs learn by associating cues with behaviours and consequences, eg: person moves their hand upward; the dog is prompted to place his bottom on the floor; the person give them a treat. Result! This behaviour will be repeated because it was worthwhile for the dog. To add a verbal cue to that sequence of events, say the cue 'sit' at the beginning, before the hand signal, which then prompts bottom placement and the ultimate reward. Dogs begin to anticipate that the sound you make ('sit') at the start of the sequence leads to your hand signal, which gives them the opportunity to place their bottom on the floor and receive a treat. Soon, they will start to sit as you say the cue, not needing to wait for the hand signal to prompt the behaviour. This pattern is known as New Cue: Old Cue, and, in this example, the Old Cue is the hand signal, and the verbal 'sit' is the New Cue.

Our dogs can learn almost anything as a cue using the same approach.

Want to teach your dog to read? Print the word 'SIT' on a piece of paper. Use the principle of New Cue: Old Cue, and hold up the paper in front of your dog; then say 'sit' and reward him once he sits. Repeat, always ensuring that you hold up the paper before you say 'sit', and reward his sit. You will see that he begins to anticipate you saying 'sit' after you have held up the piece of paper, and will sit before you say the word. With a little more repetition, he will sit just as reliably when you only hold up the paper as he does for the verbal cue.

Expand his 'reading' vocabulary in the same way with other familiar behaviours. Ensure that the printed words are easy for him to differentiate: obviously, he isn't learning to read as we understand it, but he is learning to differentiate the word shapes and lengths on the paper.

Want to use different scents as cues for different

Learning to 'Read,' or at least applying the New Cue: Old Cue format to sit when shown the 'SIT' card. She's a very clever pupil.

behaviours ... for example, chamomile means 'lie down'? Old Cue (down); New Cue (chamomile tea bag). Present the chamomile, then give the verbal 'down' cue. Reward and repeat as above.

Your dog's learning is limited only by your imagination for dreaming up useful, funny or impressive cues for his favourite behaviours.

5 Goodbye, my friend

Becoming increasingly aware of our dog's mortality is upsetting, but having a plan, or at least an idea, of the various options available when faced with the final decisions can help make a stressful and emotional time a little easier, especially in an emergency. We also need to remember to take time out to care for ourselves during this period, so we have the emotional energy to find the moments of joy amongst the more difficult times, and to take the difficult decisions when required.

Self-care is a very personal thing; what I find relaxing (reading, swimming, sitting outside under big skies and taking deep breaths) may be uninspiring to you. However, practising gratitude has significant evidence to suggest it is effective for most people. It doesn't have to be profound: just take a few moments to give thanks for three positive things you've experienced that day ... for example, the train guard who helped me off the train with my luggage and two dogs; the blue skies

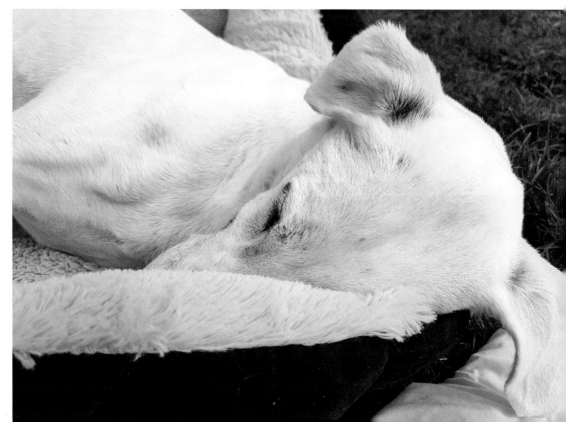

Cooper's final days were filled with love, peace, joy, and friendship. We were lucky: despite losing him way too soon and too fast, we were able to choose every detail of his passing.

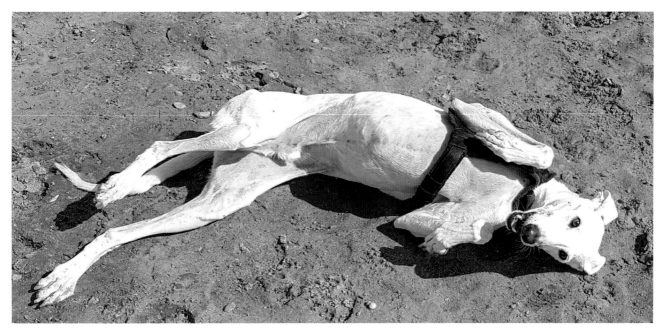

His last trip to the beach: Cooper still had the energy to enjoy a roll in the sand.

and cool winds as I walked; the fresh mint outside my door for my regular cups of tea. Whatever you can find to be grateful for can help give you the inner strength to cope with the days and weeks ahead.

It is also worth taking stock each day, and giving yourself a rating for how well, or not, you're coping. Rate your state out of 10, and make sure you act to change things (get more help, tell a friend how you're feeling, ask your vet for advice) if you're giving yourself consistently low scores.

We can do a similar thing for our dogs, and use a quality of life scale (QOLS) to help us assess their progress more objectively. How we feel about our dog's quality of life will vary depending on how we ourselves are feeling that day, so completing the scale regularly (daily, weekly, every few weeks) will enable us to spot any pattern or trend over time. An occasional low score can be explained or ignored, but a downward trend may suggest more medication is needed, or that things are approaching the point where you will have to make the final choices for your dog. There is no score that recommends euthanasia because your dog, your family, your situation is unique; but a lower score may prompt you to

consider your options, especially if your dog scores particularly low in one or two areas, despite a higher score overall.

The JOURNEYS Quality of Life Scale is included as Appendix 2.

If you are at the stage where you are completing the QOLS on a regular basis, it is probably also time to consider if there are any particular memorials or memories you want to preserve with your dog. You may want to take him on one final trip to go back to his favourite beach, or have some wonderful photos or a painting done while he is still relatively mobile, happy and healthy. There are so many options for beautiful memorials these days, from pawprint jewellery or a tattoo to a plant in your garden. Or you may choose to do nothing outwardly, because your own memories are enough. There's no judgement over grief, or at least there shouldn't be.

Now is also the time to ask yourself, and your family, how you want to prepare for euthanasia, and where you would like it to take place. Many vets will come to your home for this, and you – and your dog – may find this a more comforting option than going to the surgery. Or you may find the prospect of your beloved dog dying at home too much to bear, preferring that

Cooper loved to steal the pears from the tree in the garden, and showed all the other dogs how to pick directly from the branches, instead of waiting for the fruit to fall. I am extremely grateful to Simone from Nellie Doodles, who created this beautiful drawing for us.

particular memory to occur somewhere you don't have to see on a daily basis. Speak to your vet about your options, and agree between yourselves what you want to happen when the time comes. A description of the procedure for euthanasia is included in Appendix 3. Whether or not you choose to stay with your dog while he is euthanised is your decision, and should take into account what is best for you both.

You also have choices over what can happen to your dog after he has died. When I was a child we buried our animals in the garden, but these days it is much more common to have them cremated. Investigate the cremation options in your area: your vet will have a list, or do your own research on the internet. You can usually choose to cremate your dog individually and have his ashes returned to you either in a tube for scattering, or a memorial urn for you to keep. You can also opt for a joint cremation where you receive a portion of the ashes from the procedure. Your vet will offer to keep your dog's body until he is collected by the crematorium, or you can keep your dog at home and take him to be cremated yourself. There is a description of what happens to your dog's body after he dies in Appendix 3.

You may be concerned about how other dogs or animals in the family will react once your older dog has gone. It is often advised to allow other animals the opportunity to interact with his body, in the belief that this helps them understand he is not coming back. Again, this is a personal choice for you to make based on what's best for you and your animals. You can read how my dogs reacted in Appendix 3. Your dogs may well be subdued during the days and weeks following the loss of your oldie; this may be their own grief and loss, or they may be reflecting your sadness. Either way, take comfort from their presence, and allow everyone to grieve in their own way.

There is no right or wrong way to behave following the loss of a much-loved dog. Some people will find the house too empty, and seek to give another dog a loving home as quickly as possible. Others will wait months or years until the pain has

▸▸ *Cooper never missed a chance to steal the show, and grab some extra attention from the person with the camera.*

OLDER DOG? NO WORRIES!

reduced enough to contemplate doing it all again. Losing a dog is just as painful as, if not more so, losing a close friend or family member, and often grief will hit you when you least expect it. Let the tears flow when you need to: over time, you will be able to look at photos and remember the good times, and a piece of your heart will always belong to your much-missed dog.

Precious moments before the vet arrived. Cooper slept much of his final 24 hours.

On his last day, my younger dogs kept Cooper company, and he appeared to welcome their presence. They wouldn't normally have slept this close to him.

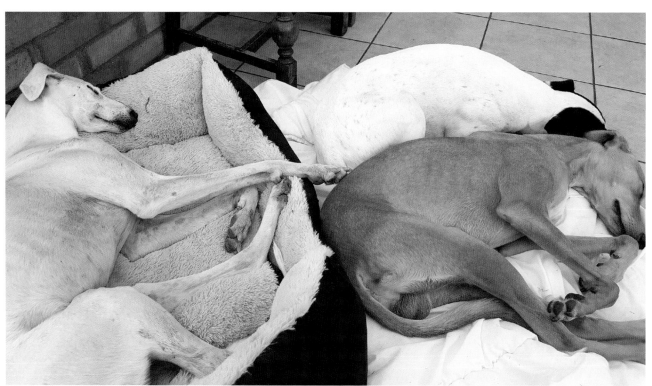

Appendix 1: Introducing scent activities

This is a very quick introduction to two different forms of scent-based activities to get you started. For a deeper understanding, and more ideas on how to develop your skills, find a local reward-based trainer who offers Talking Dogs Scentwork® classes or tracking classes for pet dogs.

Introducing scentwork

Your dog can search for cheese, or catnip-scented toys, whichever will be more fun for him. The description below assumes catnip as the default, and will give instructions for cheese only where it varies from using catnip-scented toys.

Place a tablespoon of dried catnip, wrapped in kitchen towel, in the bottom of an airtight tin, and add a few small dog

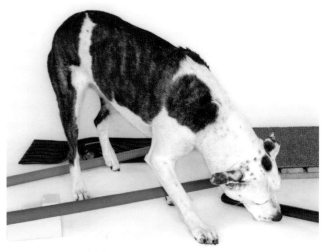

Riley is sniffing for hidden, catnip-scented mice. There is one under the mat behind her left hind leg ...

My scentwork tin. The catnip is folded inside the kitchen towel on the left. I collect small toys, or bits of larger toys, for the tin. Wash and dry after each use, before putting back into the tin.

toys on top. Leave for 24 hours for the catnip scent to penetrate the dog toys (you will later hide the scented toys for your dog to find and play with).

If using cheese, cut into small chunks; your dog will find and eat the cheese for his reward.

Let your dog watch you throw a toy a few feet away from him, then encourage him to get the toy using the cue 'Go sniff!'. When he reaches the toy, give him verbal encouragement to pick it up, if he doesn't do this of his own accord. Play with him once he has the toy. For dogs who like cheese, let him eat the cheese when he finds it, and give lots of verbal encouragement.

... and another in the top of a cone.

After a couple of repetitions where he sees you throw a toy, this time place a toy slightly out of his sight and use your Go sniff! cue to encourage him to find it.

Play with him as before when he finds the toy. Celebrate with your cheese-eating dog.

As he gains confidence, hide the toy out of sight (under the flap of a cardboard box, behind a chair, tucked in a bookcase, etc), and, if necessary, walk around the room with him to encourage him to search. Don't get in his way – he's doing the searching, not you – and resist the temptation to point the toy out to him. If he needs help, encourage him into the approximate area by tapping objects in that space, and lifting things to encourage him to sniff inside/underneath as applicable.

Use a range of different objects in which to hide the toys, including boxes, cones, platforms, bags, envelopes, old clothes or shoes – anything you have lying around, in fact.

You can work on- or off-lead, inside or outside, wherever you want, but ensure the area is safe, with nothing that could injure or harm your dog whilst he is searching.

Based on the methods of Talking Dogs Scentwork®. Any errors are my own.

Let your dog play with the toy when she finds it. Ideally, play outside of the search area, and have a game with her: sometimes, however, her joy at finding the toy takes over!

Freddie is sniffing a line of boxes for his cheese. The cheese is inside a closed box, and he is on a loose lead, checking out each box in turn ...

... and closer inspection reveals the correct box. He needed human help to open the box so he could eat the cheese, but he made the choice without human interference. When your dog is paying attention to one area or object, take a step back and ask if he has found something. If your dog stays with the object, step in and help him gain access if necessary.

Introducing tracking

Tracking is a skill in which dogs use their nose to follow human footprints, whilst wearing a harness and line to guide a person to the end of the trail.

You will need a harness that does not tighten around your dog's neck or chest when he pulls, and a lead or longline that is approximately 5m long. You will also need a pole – an

▸▸ *Freddie is checking out the pile of treats at the start of the trail, that leads out in front of him. His owner is staying back to let Freddie use his nose without interference.*

OLDER DOG? NO WORRIES!

electric fence post, a bamboo cane or similar – that is light enough to carry, and which you can easily push into the ground.

Start by laying a trail directly into the wind, ideally on grass, while your dog watches but doesn't join you. This can be easier to do with two people: one can hold your dog whilst the other lays the trail.

The pole is your starting point. Push this into the ground, and stand directly in front of it, facing into the wind. Place several treats at the base of the pole where you are standing.

Lay a trail by walking a straight line into the wind, placing one foot directly in front of the other, and laying a treat at every other footstep as you walk forward.

After approximately 10m, place a larger pile of treats on your last footstep, then take a giant step or leap to one side and walk back to your dog, taking care to keep away from the trail you have just laid.

Attach the long line to your dog's harness and walk towards the starting pole. As you reach the pole use the cue 'Go track,' and wait for your dog to discover the treats at the base of the pole. With you and your dog facing directly down the track you have laid, let him have approximately 2m of line, and make sure you always stay behind your dog with a light tension on the line. He can pull you along the track, but not run.

If your dog has got the scent of the treats along the trail, and is pulling you in the correct direction, walk behind him down the trail. If he goes off to one side or the other, wait until he comes back onto the trail, then move forward when he picks up the scent again.

When he reaches the final pile of treats, remember to tell him how fabulous he is, and remove the harness/unclip the line so he knows he's no longer tracking. If you walk back to the start of the trail, take care to do so over ground where you don't want

On the trail. When tracking, dogs will primarily use ground scent, keeping their noses low and picking up the scent of crushed vegetation, dropped human skin cells, and any other relevant scents.

to lay another trail – use the same route as the trail layer took on their way back.

Some dogs may prefer to find a toy at the end; just make sure it's hidden in a dip or behind a plant on the trail so that he doesn't just see it and head directly to it: he's supposed to be using his nose and not his eyes for this game.

As his and your skills increase, you can –

- reduce the number of treats on the trail so that he is tracking footsteps only
- lay longer trails
- wait for longer periods between laying the trail and starting to follow it so that the scent ages
- add in turns or different surfaces
- leave articles on the trail for your dog to find

To get the most out of scentwork it is worth going to classes or workshops where a skilled trainer can develop your skills, and ensure you are working with your dog in the best possible way.

Appendix 2: JOURNEYS quality of life scale

© Dr Katie Hilst, DVM
web: https://journeyspet.com/quality-of-life-scale-pets/

The JOURNEYS scale addresses eight variables you can use to help determine your dog's quality of life. Many people will use the scale daily or weekly to gauge how their dog is doing, and compare the results to see how well he has been over time. Other family members can use the scale independently and compare results. These discussions can be helpful in getting consensus on how to move forward.

For each variable there is an assigned maximum value of 10 points, with an example given as guidance for scores of 1, 5 and 10. Use your judgement to decide how your animal scores.

Example: E – Eating and drinking, if your dog 'only eats treats' you may assign a value of 2 or 3: higher than 1, which is not eating at all, but lower than 5 which is eating slightly less of his usual than he would normally.

J – Jumping or mobility

1pt

Your dog cannot walk or stand without assistance.

5pt

Your dog can move around as long as he has his pain medication. He can manage about half the activities he did when he was healthier, or can go about half the usual distance on a walk, or spend half the time doing the activities (chasing a Frisbee, swimming, hunting) he used to do.

10pt

Your dog is fully active and enjoying all of his usual activities.

O – Ouch or pain

1pt

Your dog seems in pain (whining, crying, not willing to move) even when taking pain medication. Note: many animals will hide pain or weakness as a survival trait.

5pt

Your dog is on pain medication, which is helping at least 75% of the time.

10pt

Your dog is pain-free.

U – Uncertainty and Understanding

1pt

Your dog has a diagnosis (medical condition) that cannot be predicted. You may not understand the diagnosis, or the condition may be prone to sudden, catastrophic events.

5pt

Your dog has a medical condition that can change over time, is currently stable, and you are able to monitor it (with the help of your veterinarian), and make adjustments when necessary. You understand what to watch for, the treatment plan, and when your dog needs medical attention.

10pt

Your dog is happy and healthy; there are no medical issues beyond routine preventative care.

R – Respiration or breathing

1pt
Your dog has severe episodes of breathing difficulties, coughing or open mouth breathing. He is not eating or drinking in his effort to breathe. Seek immediate medical attention for your animal.

5pt
Your dog has occasional bouts of coughing, wheezing, or exercise intolerance. These are short (under 2 minutes), he is on medication from your veterinarian, which can be adjusted to help.

10pt
Your dog has no coughing, wheezing, or exercise intolerance.

N – Neatness or hygiene

1pt
Your dog spends time laying in his urine and/or faeces; he may be unable to control elimination, or be unable to move after elimination. Your dog may have an external tumour or mass that is bleeding, foul-smelling, and infected, and you are unable to keep it clean, and/or bandaged. Your animal may have pressure sores (bed sores) from lying down and being unable to move.

5pt
Your dog may need assistance to urinate/defecate but does not spend time lying in his waste. He is able to hold his bladder/bowels until assistance is available. He may have an external tumoor or mass, but this can be kept clean, and/or bandaged, and is not infected. He grooms himself, but may need assistance in some areas (rear end, for example).

10pt
Your dog can urinate, defecate, and groom himself without assistance. He has no medical issues that are causing him to have a bad odour. You can provide any care issues to address his hygiene (baths, trip to the groomer, anal gland expression, teeth cleaning, etc)

E – Eating and drinking

1pt
Your dog is refusing food and water. He may be vomiting or having diarrhoea (or both). He may be nauseous.

5pt
Your dog is eating more slowly, and is not as interested in food as he used to be. He may go back several times before he finishes a meal. He is eating slightly less than usual, but eating his regular food.

10pt
Your dog is eating and drinking normally.

Y – You

1pt
You are constantly worried about your dog, and may not understand what is happening to him. You feel overwhelmed and stressed trying to provide for his needs, and may feel unable to provide for these physically, emotionally or financially. You may be worried about how he will fare when you are away on an upcoming trip. There may be tension in the family and disagreement on how to proceed.

5pt
You understand your dog's condition, and are able, with some effort, to meet his needs. You may have concerns, but they are manageable.

10pt
You are easily able to meet your dog's needs, and not worried about any aspect of his care.

S – Social ability

1pt
Your dog does not spend time with the family. He may hide, become irritable or snippy if bothered. Some dogs who do not usually enjoy being petted may not seem to care if they are

petted. Perhaps your dog is physically unable to get to the room where he usually spends time with others.

5pt
Your dog spends at least half the time with the family. He is not irritable or snippy, and happily greets you when you come home.

10pt
Your dog enjoys you, the family, and others (including other animals he may know), greets you at the door when you arrive home, and seeks out company.

Sometimes, after a discussion, people realize that their dog is, in fact, enjoying life, and they still have time left with him. Other times, people realize that their dog is actually suffering more than they thought, so they choose the final act of caring. In either case, the JOURNEYS scale is meant to get you thinking and considering the factors that affect your dog's happiness, sense of well-being, and quality of life.

When it comes to scoring, there are no hard and fast rules, although, in general, a higher score is better.

- A score of 80 is a happy, healthy dog!
- A score of 8 is a dog who is suffering.

A low score on any one of the variables *may* be a reason to consider euthanasia if the score cannot be raised. An example of this would be a dog who is still eating and drinking, but who has pain that cannot be controlled, even with pain medication. Another example is a dog who scores 10 in social contexts, but who has difficulty breathing. We purposely do not give a score at which euthanasia should be considered, because every dog, situation, and family is unique. You should consider your animal's personality and needs when making the difficult decision.

Appendix 3: Cooper's final days

Written October 2018, a month after he died. Previously published on the Developing Dogs Blog

Cooper was diagnosed with an aggressive cancer, which had spread to secondary tumours in his lymph nodes, a few days after our summer holiday. He had a marvellous summer – one of his favourite things was to sunbathe, and he had plenty of opportunities that year. Our holiday was also perfect; the only shadow being the lumps we found in his neck while we were there. As he was otherwise, apparently, fit and well, playful and happy, we decided not to get them investigated while we were in France, but waited until his lovely regular vet could check him out. We braced ourselves for a diagnosis of lymphoma – something which may have given us a few more months at least. The truth, that we were facing weeks at most, was almost devastating. We didn't know how long we'd have, but when your vet gives you only three weeks worth of tablets then you know it isn't long.

There were times during those next few days when it almost felt like I'd never be able to breathe again. Cooper was my special boy, my favourite (it's okay to have favourites); I thought he would have several more years before having to face this. I cried myself to sleep on several occasions, while planning all the things I wanted to squeeze in to his final few days while he was still well enough.

For the first week, I was sometimes able to persuade myself we could have months. He didn't seem any different – playing with the younger dogs; enjoying his walks – but gradually he slowed down, his breathing became more laboured, and ten days after his diagnosis he slumped. Suddenly, we couldn't control his pain and he was miserable. Frantic email to the vet – because I couldn't talk without sobbing on the phone – and we had a plan to increase his meds to block his pain. But on that Saturday morning I knew his time was short.

We took him to the beach. it was a glorious day, and we have some amazing photos to treasure from it. He got a burger in the pub for his dinner. The following day we took him away for the night, leaving the others with a friend to look after, and it was there we agreed that his time had come. He got steak in

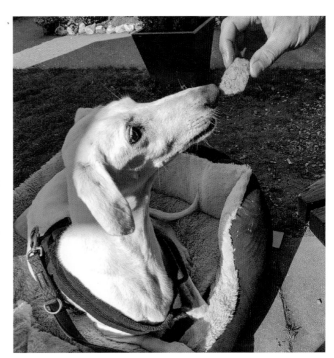

Cooked breakfast from the pub we stayed in for Cooper's final trip away. The steroids helped manage his pain and keep his appetite strong.

the restaurant, a cooked breakfast the following morning, and was spoiled by everyone he met. He managed a short walk on the beach the next morning – still greeting strangers with his trademark grin, and lapping up the attention. We came home, special friends came to say goodbye, and then we wrapped ourselves up for his final 24 hours. He had been a big part of the lives of lots of people, but we didn't want to share him any more.

While we were away we had planned what would happen. We asked our vet to come to the house to put him to sleep. We planned that we would keep his body at home, and take him to be cremated privately the following day. It was important to me that I took care of him, even after he'd gone.

Taking time to say goodbye. It was important to us that we got to spend time as just the three of us, without other dogs or other people.

His final day was warm and sunny. He pottered in the field and the garden, he slept (accompanied by the two youngsters who seemed to know he wasn't his usual self), and spent the afternoon in the garden asleep on his bed.

We shut the younger dogs away so that when the vet arrived it remained peaceful and calm. Cooper got up to greet him, but then went back to his bed. Your vet will be accompanied by a vet nurse to assist with the procedure, and hold your dog if you're unable to do so.

When your dog is put to sleep, the vet will shave a small portion of skin so the needle goes in cleanly. As he slowly administers the drug, your dog will slip away. There may be some involuntary movement, but your dog won't feel anything. His eyes stay open; he may urinate or defecate as muscles relax. We believe that a dog's hearing is the last thing to go, so we cuddled him as he slipped away, and told him what a gorgeous and handsome boy he was, and that it was okay to go. Afterwards we sobbed, and sobbed. Riley had watched much of what was going on, but didn't want to investigate his body at all.

When the vet had left, and we were ready, we let the younger dogs out. They came racing out to where Cooper was lying, apparently asleep in his bed. Buzz hesitated as he approached, and curved round him, taking a few moments to check him out. Woody, on the other hand, ran straight up, stood on Cooper's body, and shoved his face into Paddy's face. He knew instantly that Cooper wasn't Cooper any more; he would never have done that while he was alive. Even in that moment of pure grief, I was able to marvel at a dog's ability to read another dog.

Cooper stayed at home, lying on a bed, covered with a blanket but with his head out because I couldn't bear to cover it, until I took him to be cremated the following day. I checked on him several times that night. He gradually went stiff and cold to the touch. Small amounts of blood escaped from his mouth, and urine soaked the towel we'd placed between his legs. In Cooper's case there was a smell, though this isn't always noticeable.

The crematorium was calm and peaceful. I took him first thing in the morning, and knew they would take care of his final moments. He was lifted on to a table alongside the incinerator and I kissed him one last time. When I collected his ashes three hours later, they were still warm. He's now home.

I still cry for him most days; there's a huge hole where he used to be. I am getting used to working without him, and Buzz is stepping up to do Cooper's jobs differently, but successfully. The dogs are settling down, without the great white goof joining in when the youngsters play. Riley is a bit unsettled and wants to come along for the ride when we go out, when previously she'd have preferred to stay home. But we're getting there, in our own way.

We were lucky; we had time to plan. And we did what felt right for us. It doesn't matter what you do – everyone is different – but, if you can, decide now what you want to happen. You can always change your mind, and circumstances may force that on you, but exploring the options in advance means you face fewer decisions to make at a time when the world is spinning, and it feels like life is going on without you.

Index

Rogues' Gallery: The Cast

Riley

Meg

Tess

Cooper

Billy

Bacon

Poppy

Pippin

Diva

Freddie

Meeting a dog's physical, mental, and emotional needs during a period of limited mobility can help reduce the possibility of future behaviour problems, alleviate some of the stress of caring for a less active canine, and help aid recovery.

Encouraging owners to reflect upon, and take into account, their dog's individual requirements in advance of surgery or other lifestyle-limiting event, this book also contains information and advice about appropriate activities that owners can introduce to their dog's daily routine whilst walks are limited.

Paperback • 205x205mm • 72 pages • 120 colour images • ISBN: 9781787115057 • £13.99*

Prices subject to change • p&p extra